WORK LOG

PERSONAL INFORMATION

Personal

Name:

Address:

Tel.:

Email:

Cellular:

Business

Tel.:

Email:

Cellular:

Contact Names	Telephone	Email
Accountant:		
Bank:		
Club:		
Dentist:		
Doctor:		
Garage:		
Insurance Agent:		
Lawyer:		
Mechanic:		
Other:		

Details — Expires

Details	Expires
Blood Type:	
Driver's License:	
Health Insurance:	
Insurance Policies:	
Club Membership:	
Passport No.:	

Important Dates

Date	Name	Event

WORK LOG PLANNER

Name	January	February	March	April	May

WORK LOG PLANNER

June	July	August	September	October	November	December

BUDGET

Month/Year: _____   **Month/Year:** _____

Month/Year: _____   **Month/Year:** _____

Month/Year: _____   **Month/Year:** _____

BUDGET

Month/Year: **Month/Year:**

Month/Year: **Month/Year:**

Month/Year: **Month/Year:**

DIRECTORY

NAME	ADDRESS	TELEPHONE

DIRECTORY

CELLULAR	FAX	EMAIL

THE IMPORTANCE OF KEEPING A WORK LOG

Do you have a job?

Do you keep a record of what you do on your job?

Did you know that setting aside 15 minutes at the end of the day to record in a Work Log and reflect on your day can boost your efficiency and thus impact your career success?

In addition to this, a Work Log is a record of actions, events, accomplishments, and incidences. Record activities in your Work Log hourly, daily, weekly or even monthly. But why is it important to keep a Work Log? A Work Log:

a. Helps to keep a record of your daily activities such as clocking in and clocking out times

b. Helps to record tasks that you accomplish throughout the day,

c. Can be used to keep only important information, without too much detail

d. Allows you to record when and who gives you a task or to whom you give a task,

e. Allows for an easier preparation of reports by referring to your Work Log,

f. Can be used to record sick days, absences, lunch time and even your salary,

g. Provides a hard copy in your own handwriting,

h. Assists you in providing legal evidence in case of legal proceedings against you,

Choose from our wide selection of Work Logs and customize it to match your needs.

Please leave a review or send us a copy of your customized Work Log to *keyworklogs@gmail.com* so that we can improve our Work Logs to serve you better.

Work Log size 8.5 x 11 inches (Simply click on the name Key Work Logs beside the word Author to see Work Logs in other sizes)

Made in the USA
Monee, IL
30 September 2021